I0470630

TRUTH, PERCEPTION, AND CONSEQUENCES

The Proteus Monograph Series

Volume 1, Issue 1
November 2007

The Proteus Management Group

The National Intelligence University, Office of the Director of National Intelligence and the Center for Strategic Leadership, U.S. Army War College established the Proteus Management Group (PMG) to focus on examining uncertainty, enhancing creativity, gaining foresight, and developing critical analytical and decision-making processes to effectively provide insight and knowledge to future complex national security, military, and intelligence challenges. The Group is also closely associated with Proteus and Foresight Canada.

The Proteus Monograph Series Fellows Program

The Proteus Monograph Series Fellows Program (PMSFP) funds research by scholars who are not affiliated with the U.S. Government and military. These scholars can be located either in the United States or abroad. This program is designed to support the Proteus Management Group's charter to promote further discourse, study, and research focusing on the refinement, development, and application of new and emerging "futures" concepts, methods, processes, and scenarios. The overarching goal of this effort is to provide value-added and relevant commentary that will assist strategic and high-operational level decision makers, planners, and analysts with "outside the box" considerations and critical analysis of national, military and intelligence issues within the Joint, Interagency, Intergovernmental, and Multinational (JIIM) environment.

Truth, Perception, and Consequences

by

Christine A. R. MacNulty, FRSA

PRESIDENT & CEO, APPLIED FUTURES, INC.

Ms. Christine MacNulty has more than thirty-five years experience in long-term strategic planning for cultural change, technology forecasting, and technology assessment. She consults at to the most senior levels within the Department of Defense. Her current DoD projects bring together her knowledge of strategy, cultures, and cognition to help in understanding our adversaries in order to develop non-traditional operations, information operations, and strategic communications. For her contribution to British industry, she was elected a Fellow of the prestigious Royal Society of Arts, Manufactures and Commerce. She has authored numerous papers and is a very popular conference speaker. She has co-authored of two books: Industrial Applications of Technology Forecasting, and, Network-Centric Operations: Translating Principles into Practice (to be published in 2007). She is the founding President and CEO of Applied Futures, a consultancy based near Washington, D.C.

TABLE OF CONTENTS

ABSTRACT

Today we, in the United States, tend to regard Sun Tzu's maxim "If you know the enemy and know yourself, you need not fear the result of a hundred battles" as being about the physical capabilities, C4ISR (Command, Control, Communications, Computers, Intelligence, Surveillance and Reconnaissance), weapons, and warfighting capability of our own forces and those of the adversary. Given the rest of *The Art of War,* and its focus on preventing such actions, it is likely that Sun Tzu was referring much more to the understanding of the psyche than to the material aspects of warfare.

In the last few years, we have seen many commentaries from former Secretary Rumsfeld and others that indicate that we are not winning the war of ideas. In addition, it has been clear that we have had some significant failures in intelligence. Some recent conversations indicate that we may be failing in the information/ideas war because we, have not taken Sun Tzu's lessons to heart.

While my concerns are strategic—because I do not believe we have been able to craft real strategies for intelligence, Information Operations (IO), and other non-kinetic operations—we have not even been doing well at the tactical end either. For instance, in an online journal, Major Bill Edmonds, U.S. Army Special Forces, said "I have slowly come to understand that if we are to succeed in Iraq, we must either change the way we perceive and treat those we want to help, or we must disengage the great percentage of our military from the population."

Intelligence, IO, Psychological and Influence Operations, and Strategic Communication (SC) all require a deep understanding of our adversaries and, in some cases, our allies. They also require a deep understanding of our own cultural biases, so that we see as clearly as possible while looking through our own cultural lens. While there may be a single "truth" out there, most of what we "see" is perception, not truth; and most of what the adversary sees is also perception, not truth. We have done some things well and have achieved some of our objectives. We have sometimes achieved first order results at the expense of longer-term strategic goals, and in some cases, we have failed to achieve our desired results and have created unintended consequences. Many of these failures have happened because of

- lack of a systems perspective and the ability to see patterns on a strategic scale;
- lack of understanding of the cultural context;

- lack of understanding of some key cognitive dimensions of adversary decision making;
- lack of understanding of the cultural-cognitive relationships;
- lack of understanding of the nature of and complexity of the systems with which we're dealing.

This monograph focuses on the key elements of understanding cultures: the stories our cultures develop to make sense and meaning from the world; the metaphors we use and how those help to frame perception; and the nature and determinants of our mind-sets. I have illustrated all of these with historical examples from the last fifty years.

Finally, the monograph outlines a limited number of critical cultural-cognitive dimensions that can be used to evaluate an adversary—including his values and motivations—so that we can anticipate his actions and better determine how to influence them. It also recommends a checklist of other things that we can do to enhance our cultural awareness and understanding, and it suggests the kinds of further research that are needed.

TRUTH, PERCEPTION, AND CONSEQUENCES

Introduction

Today we tend to regard Sun Tzu's maxim "If you know the enemy and know yourself, you need not fear the result of a hundred battles" as being about knowledge of the physical disposition, C4ISR (Command, Control, Communication, Computers, Intelligence, Surveillance and Reconnaissance), weapons, and warfighting capabilities of our own forces and those of the adversary. Given the rest of *The Art of War,* and its focus on prevention, it is likely that Sun Tzu was referring much more to the understanding of the psychological (culture and cognition) than to the material aspects of warfare. This implies an understanding of what the different parties are seeing, how they understand the situation, what they are thinking about it, and what they are thinking of doing about it. Yet we do not seem to be good at that:

> "We are not winning the war of ideas." –Secretary Rumsfeld

> "We are not winning the IO [Information Operations] War." –COL Randy Gangle, USMC (Ret), USMC Center for Emerging Threats & Opportunities[1]

> "Despite its own multi-cultural nature, the (U.S.) Army was not culturally attuned to the environment (in Iraq)." –Brigadier Aylwin-Foster, British Army[2]

> "US Army personnel instinctively turned to technology to solve problems." –Brigadier Aylwin-Foster[3]

Most people accept the four domains of warfare articulated in many publications from the Office of the Secretary of Defense (OSD) – the Office of Force Transformation and the Command & Control Research Program. In most cases, those four domains—Societal, Cognitive, Informational and Physical—are represented as discrete layers, as in a cake. In reality they are nested, with the Societal (or Cultural) domain underpinning the other three; and the Cognitive domain underpinning the Informational and Physical. This is important, as it illustrates the fact that our decision-making processes are affected by our culture. The information that we choose to collect is determined by our culture and the ways our minds work, and even the

---

1. Colonel Randy Gangle, USMC (Ret), *Aviation Week and Space Technology* August 9 2004.

2. Brigadier Aylwin-Foster, British Army, *Military Review* November 1 2005.

3. Ibid.

weaponry and systems we develop are products of our cultures and thoughts. We have ideas about truth, and we are strong upholders of "the truth," but what is really true? Is there an absolute "truth," or to what degree is truth a function of our culture and our perception?

Which is true?

- Is a billion 10^9 or 10^{12}? and is a trillion 10^{12} or 10^{18}?

- Was Exxon's profit last year $83,809,000 or $39,500,000?

- Is the United States the "land of the free and the home of the brave" or the "Great Satan"?

The answer is that *both* choices are correct for each question. Let us think about them and their implications in more detail.

In the United States a billion is 10^9 and a trillion is 10^{12}. In the United Kingdom and Europe a billion is 10^{12} and a trillion is 10^{18}. While there are probably some economists in both countries who understand the difference, think about the response of even well-educated Britons and Europeans to headlines that discuss the U.S. budget, Federal deficit, or trade deficit. No wonder they regard us as greedy and feckless, when they understand those deficits to be a billion or a trillion times more than they are in our reality.

The figures for Exxon's profit are for gross profit and net profit respectively. But the headlines we see in most media, which often quote the gross profit, do not explain that. Nor do they explain that the costs of exploration and extraction of oil are increasing, and that without profits of that size, Exxon could not continue to provide the oil that we require.

While we see the United States as the land of the free, the home of the brave, and the land of opportunity, Iranians of a radical turn of mind see our television that suggests that we engage in sexual promiscuity, drug abuse, and other "sins," and they truly believe that we are offering great temptation to young people—hence the Great Satan.

As we shall see, differences in perception of this sort can be of critical importance.

The Importance of Knowing the Enemy

There are two major reasons for knowing the enemy:

- To be able to anticipate his actions—by understanding *why*, not just what and how

- To be able to influence his actions—or otherwise communicate with him

Anticipation of actions cannot be done effectively by extrapolating behavior. It requires an understanding of the motivations and mind-sets of the adversary. This will be discussed at length later in this monograph. Since communications permeate all interactions between people, we will consider that first.

St. Augustine identified the problem of communication in *De Magistro,* when he said "If I am given a sign and know nothing of which it is the sign, it can teach me nothing. If I know the thing, what do I learn from the sign?"[4] In other words, signs (words, symbols) by themselves do not convey meaning. We infer meaning from the context, the individual communicator, and the style, intonation, and other characteristics of what is being conveyed.

Information Operations, Psychological, and Influence Operations all require extremely effective communications from us to our adversaries. Intelligence also requires effective communication and understanding. From its Old French origin, communication means to impart, to share with others. This requires an understanding of those others in order to ensure that what we are trying to say is understood by them. We should not assume that everyone gives the same interpretation to the words, phrases, and symbols we use. We in the West, and particularly in the United States, tend to believe that there is only one truth and that others see and understand as we do. In the Armed Forces, this is known as "mirror-imaging"; in anthropology it is known as ethno-centrism. While we have had successes in Iraq and Afghanistan, it has been clear to many people, including the former Secretary of Defense, that we are not winning the war of ideas. We have not been as successful at information operations as some of our adversaries. This has been caused, in part, by our Rules of Engagement (which our adversaries do not have), but it has also been caused by our lack of understanding of the cultures of our adversaries and those who support them. Their perceptions and their "truth" are very different from ours.

There is not even a single definition of truth about which most philosophers agree. The word is derived from the Anglo-Saxon word *treow* (true), which means faithfulness in the quality of being accurate. Yet there are five theories of truth, ranging from a perspective of absolute, objective truth to a truth that is constructed by social processes or consensus. Lakoff and Johnson suggest that there is no such thing as objective truth, and that the belief in such a truth is not only mistaken but socially and politically dangerous.[5] We need to be aware of these perspectives and their implications

---

4. Jorge J. E. Garcia, *Old Wine in New Skins* (Milwaukee: Marquette University Press, 2003), 18.

5. George Lakoff and Mark Johnson, *Metaphors We Live by* (Chicago: University of Chicago, 1980), 159-162.

as we attempt to achieve genuine communication amongst ourselves, with our allies, with those who are neutral, and with our adversaries. These authors also pose a question about what it takes to understand a simple sentence.[6] For instance, in the sentence: "John fired the gun at Harry," we have to understand that John and Harry are proper names; we need to understand what a gun is and what it means to fire a gun and to fire it at someone. We also need to understand the implications of firing a gun at someone; in other words, Harry might be wounded or killed. If all this needs to be understood if we are using the English language, what more do we need to do when conveying Western concepts to other cultures?

Perception comes from the Latin meaning to receive or collect; it later became defined as intuitive recognition. Each society, group, and individual has its own way of perceiving, which is framed by, or colored by, its cultural background. Taking truth and perception together, we need to realize that each of us sees and understands the truth of the circumstances and events going on around us through a "lens" that is composed of the "story" in which we live. When we share those perceptions with others who have the same or a very similar "lens," we tend to find agreement. When we share perceptions with those who have a very different lens, disagreement—even to the point of believing that others are lying deliberately, which may not be the case.

Why We Need Stories to Live By

Each of us lives inside a metaphor, or story. The story is based on the events and circumstances that have created meaning in our lives—our experiences. We invent the stories in order to integrate our interpretations of those experiences into a coherent whole. We need meaning to make sense of our world, and *that meaning is created when our intellect and emotion are engaged simultaneously.* Yet, with our emphasis on science, technology, and intellect, we in the West often ignore the emotional aspects of our stories. While we have stories we tell ourselves to explain our own experiences, we also have stories we tell ourselves to explain the behavior of others, whether those be individuals, groups, or nations.

The metaphor provides a simple and easy definition of the story. "Life, liberty, and the pursuit of happiness" is one of the unalienable rights of Man as stated in the Declaration of Independence, and for many it is still the metaphor for America itself. The U.S. Armed Forces fight not just for the liberty/freedom of Americans, but for the freedom of others around the globe. Sometimes we use images as the metaphors: Uncle Sam, an authority figure representing the United States Government; John Bull—a yeoman figure

---

6. Ibid., 166-169.

representing common sense and doggedness—for the United Kingdom, and Marianne—representing Liberty, Equality and Fraternity—the symbol of the French Republic. All those symbols were generated during the 18th Century, and while the older members of those societies still know of them and understand them, the younger members and immigrants probably do not.

We probably think about metaphors as poetic imagery, yet Lakoff and Johnson have concluded that metaphors pervade all aspects of life and that our ordinary conceptual system of thinking and acting is fundamentally metaphorical in nature. As an example of this, they posit the American concept that "Time is Money" as follows: [7]

- You're *wasting* my time
- This gadget will *save* you hours
- I don't *have* the time to *give* you
- How do you *spend* your time?
- That problem *cost* me an hour
- *Put aside* some time for socializing

This Time is Money metaphor would not apply in Latin America or the Middle East. They have their own metaphors, not about Time and Money, that are meaningless to us. Lakoff and Johnson argue that, by understanding the metaphorical nature of the language used, we can gain an understanding of the metaphors, and thus the stories, that the people in a culture live by. [8]

Every major society or grouping around the world has its own story. The West shares a story that is based, to a large extent, on its Judeo-Christian heritage; yet there are differences between the Catholic and Protestant countries, especially with regard to the role of work and economics. The United States is more religious and capitalist than Europe, which is more secular and socialist. Religious groups have their stories, which cut across the stories in nation-states. Muslims are divided into Shia and Sunni, and are more or less authoritarian, depending on the particular sects in certain countries. Political persuasions, especially communism and socialism, have their stories that also span nations. Today we see the development of other movements that span nations—environmentalist movements, political movements, terrorist groups.

---

7. Ibid., 7-8.
8. Ibid., 6.

Finally we have individuals, and each individual has his or her own story. To a large extent, individuals in a particular society will have been enculturated into the beliefs and norms of that society as they were growing up. Yet even members of the same family will have slightly different stories, depending upon their experiences and their psychological tendencies.

So why is all this important?

If we hope to influence people, then we have to enable them to see that what we are saying fits with their story; or, at least, that it is not so far removed from their story that they think that what we are saying is ridiculous or that we are lying. It is the communication version of the martial arts in which we use the force generated by our opponent himself to throw him off balance. And since, as we have said earlier, stories engage both the intellect and emotion, we will do better at this kind of communication if we can include appeals to emotion in whatever we say. Indeed, the situation is even more complex than this, as we shall see later.

As an example of each person having a different story, consider this. We have all heard about several witnesses to a car accident who describe what appear to be entirely different events. Sometimes this is because they have very different physical perspectives of what happened (views blocked by people, trees, other vehicles) and sometimes because they have different beliefs. For instance, if a sports car is involved, a witness who believes that all drivers of sports cars are reckless is likely to give a negative statement, whether or not the driver was doing anything reckless at the time.

Some of us have stories that are effective in leading us to good, productive, wealth-creating lives. In the United States, Robert Allen has founded the Enlightened Wealth Institute with the aim of creating a million enlightened millionaires. His view is that millionaires benefit not only themselves, but the economy at large, and he insists on his protégés donating to charity. His philosophy is that infinite money awaits everyone who applies the principles of acquiring it—and he teaches those principles. This idea resonates with many Americans, but is completely befuddling to many Europeans whose stories are less individualistic and more collective, and who regard the making of money as somehow greedy.

If we transpose these ideas of stories to the world of international relations and conflict, where the players generally have extremely strong emotions, the possibility for misunderstandings and misinterpretations of observations become overwhelming unless we and the other parties try to "see into" each others' stories. For example, Michael McConnell tells of a meeting with a young Iranian, Mehdi, who had been tortured by the SAVAK.[9] Mehdi and his brother were convinced that, because the Shah was an enemy to Islam and wanted to denigrate Muslims, the Shia in Iran were being given frozen meat to eat that came from the dead bodies of men and women from the southern United States. He was also convinced that the Shah was the puppet of the Jews. With these and several other pieces of a story firmly in his mind, he was able to explain all manner of other actions of the Shah in such a way that he was convinced the Shah was a devil that must be fought, along with his supporters, the Jews and Americans. His story combined both logic (albeit falsehoods) and emotion, which is why it was such a powerful influence in his life. It is perhaps not easy for us to believe that someone sane could fabricate such a story and truly believe it, but that is why we need to understand people like this before we can think about communicating with and influencing them.

In the same book, McConnell describes the stories of many young people who turned to such violent political movements as the Baader-Meinhof Gang in Germany and the Red Brigade and Prima Linea in Italy.[10] Many of those in Italy came from the group known as the *non-garantiti,* young dispossessed people who had dropped through the cracks of the Italian economic miracle, and who were half-educated, unskilled, and unemployed. They saw the Maoist movements that promised workers' autonomy, factory seizures, and appropriation of property as being the way to solve these problems. And in doing so they were rebelling against their fathers' more traditional communist values. They had convinced themselves that their new story would work, even if it took violence to achieve it.

Einstein said that no problem can be solved from the same level of consciousness that created it. When we are inside our story, we tend not to recognize it and its biases, and we can certainly not recognize another's story. Too often, another's story appears to be fantasy or lies. True communication can occur only if both parties can rise above their stories and "see" them from some higher level of perspective. This requires a significant understanding of cultures, stories, and psychology/cognition. It requires making these "invisible" aspects of a culture and of a person's psychological state, visible. This is what we must do if we want to influence people effectively. The purpose of this monograph is not to explore psychology or neuroscience, but rather to provide a practical means for understanding stories, metaphors,

---

9. Michael McConnell, *Stepping Over: Personal Encounters with Young Extremists* (New York: Reader's Digest Press, 1983), 28-38.

10. Ibid.; for the Baader Meinhof Gang, see pp. 51-60; for the Prima Linea, see pp. 218-244.

and other cultural and cognitive dimensions so that we can develop tools for planning and assessing interventions of various kinds.

How can we begin to understand these stories? By "listening to the conversations" of our target audiences on the web—in chat rooms, blogs, MySpace, YouTube, terrorist websites—in videos and movies, as well as on television channels, and in coffee shops and market places. Jokes and cartoons often contain real truths about how people are feeling about things, and stories can be deduced from some of these. To get the most out of these sources, we need interpreters and translators who are current in the languages and in modern slang and colloquialisms. We need experts who understand the history, including recent events, and we need specialists in semiotics—the science of symbols, which includes symbols, signs, gestures, and intonations. These sources of information and the approaches used to understand stories can also be applied to perceptions and mind-sets, as described below.

Perception—the Act of Perceiving

There are two parts to perception—the observation of something (visual, verbal, physical) and the recognition of what it is or what it means. In Gestalt (which means "organized whole") psychology we find an approach that emphasizes the context and the way we perceive objects as well-organized patterns rather than separate components. This can sometimes result in our seeing something that is more than the sum of its parts.

(a) Gestalt Theory of Visual Perception[11]

The focal point of Gestalt theory is the idea of "grouping," or how we tend to interpret a visual field (or problem) in a certain way. There are six main factors that determine grouping:

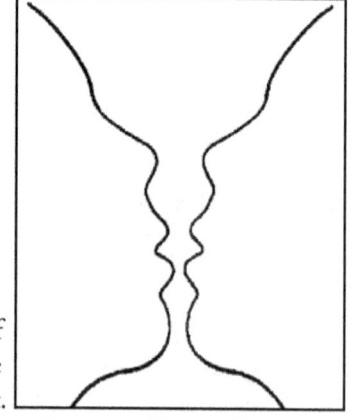

1. Proximity

2. Similarity

3. Common Fate

4. Continuation

5. Closure

6. Area

Figure 1 illustrates the gestalt idea of "figure/ground." Some people will see the vase first, while others will see the two faces first. It depends on what you see as the foreground

FIGURE 1. FIGURE/GROUND

---

11. John Powderly—derived from notes taken during his undergraduate degree program.

and what you see as the background (ground). What you see as figure and what you see as ground is dependent upon six interdependent laws, which are shown in Table 1.

1. Proximity The law of proximity states that things which are closer together will be seen as belonging together.	● ● ● ● ● ● ● ● ●
2. Similarity The law of similarity states that things which share visual characteristics such as shape, size, color, texture, value or orientation will be seen as belonging together.	□ □ □ □ □ □ □ □ □ □ □ □ □ □
3. Common Fate This law is difficult to represent here, because it is where objects *moving* in the same direction are seen as a unit.	
4. Continuation The law of continuity predicts the preference for continuous figures. We perceive this image as two crossed lines instead of four separate lines meeting at the centre.	C A O D B
5. Closure The law of closure applies when we tend to see complete figures even when part of the information is missing. For example, here we see a circle, rather than an arc; and three black circles covered by a white triangle, rather than three incomplete circles. And, in another example, we see a triangle, although no triangle has actually been drawn. Our minds react to patterns that are familiar, even though we often receive incomplete information.	
6. Area The law of area states that the smaller of two overlapping figures is perceived as figure while the larger is regarded as ground. We perceive the smaller square to be a shape on top of the other figure, as opposed to a hole in the larger shape.	

TABLE 1. SIX INTERDEPENDENT LAWS OF GESTALT PSYCHOLOGY

These perceptions are also culturally based. Some cultures are more aware of the figure, while others are more aware of the ground.

(a) Differences in Thought

The psychologist Richard Nisbett has discussed the historical and cultural differences between the West and East in his book *The Geography of Thought.*[12] He sees some of the key differences as being derived from the Confucian/Taoist heritage of the East and the Aristotelian heritage of the West. The Chinese held the view that the world was composed of a mass of substances rather than discrete objects, and hence, they had a view that life was interdependent, holistic, and harmonious. Their emphasis, therefore, was on the collective and relationships, and their view of the world was one of complexity. However, the Greek view started with individuals and individual objects and dealt with the properties of the objects. The Greeks had a much simpler view of the world and believed that it was knowable, but they failed to understand the nature of causality.[13]

In practice, these differences can give members of the different cultures entirely different perceptions of what is going on. For instance, a group of American students and a group of Japanese students were shown some videos of underwater scenes that contained fast-moving fish, slower-moving creatures, plants, bubbles, gravel, and rocks.[14] Each group was shown the video clips twice, and the participants were then asked to describe what they had seen. The Americans and Japanese both noticed the fast moving fish, but the Japanese made 60% more references to the background elements—gravel, plants, etc. On the whole, the Japanese tended to refer to the whole environment, with remarks such as "It looked like a pond," whereas the Americans referred mostly to the largest or fastest moving fish.

In another experiment, Japanese and Americans were shown still, computer-generated film clips of an airport scene and were asked to note differences in the clips. As Nesbitt anticipated, the Japanese noted many more differences in the backgrounds of the pictures, such as the airport control tower having an entirely different shape, while the Americans noticed more of the foreground changes, such as an aircraft with its landing gear up in one picture and down in the other. These examples of Eastern recognition of background and American recognition of foreground may have significant implications for visualizations and displays such as the Common (Relevant) Operational Pictures [C(R)OP]. When we and our allies view the same picture, they may not see exactly what we see and vice versa.

Nisbett also recounted examples of mass murders in China and America.[15] The Chinese attributed both murders predominantly to the context and the

---

12. Richard Nisbett, *The Geography of Thought: How Asians and Westerners Think Differently and Why* (New York: The Free Press, 2003).

13. Ibid., 8-28.

14. Ibid., 89-96.

15. Ibid., 111-117.

situational factors in which the murderers found themselves, while the Americans attributed them predominantly to the personal traits and dispositions of the murderers. Again, this illustrates the Eastern recognition of a complex system of interdependencies, and the American view of more independent, individual action.

In general, Nisbett postulates areas of difference in thought (Table 2 is derived from his work).

Area of Interest	Eastern Approach	Western Approach
Attention and perception	Environment, relationships (Background)	Objects (Foreground)
Composition of the world	Substances	Objects
Controllability of environment	Difficult	Possible
Stability versus change	Change	Stability
Explanations of events	Environment, context	Objects, individuals
Organization of world	Relationships-based	Category-based
Use of formal logical rules	Less inclined	More inclined
Application of dialectical approaches	Seek Middle Way	Correctness of one belief versus another

TABLE 2. EASTERN/WESTERN DIFFERENCES IN THOUGHT (AFTER NISBETT)

Cultural taboos also play a strong role in perception. This is illustrated by an anecdote in *The Influence of Culture on Visual Perception* by Segall, Campbell, and Herskovits.[16] During World War II, American soldiers and Algerian laborers were in sufficiently close contact that they saw each others' cleanliness habits. It is probably not surprising that the Americans viewed these impoverished Algerians, who lacked both water and soap, as filthy. What might be more surprising is that these Algerians viewed the Americans as filthy, and with even more disdain. This was because the Americans put food into their mouths with the same hand they used to control direction when urinating—something that is deeply engrained as taboo in Islamic cultures. This taboo against using the same hand for these two different purposes came from a nomadic society that lacked water, except for drinking; and in desert

16. Marshall H. Segall et al., *The Influence of Culture on Visual Perception* (New York: Bobbs-Merrill, 1966) 14.

areas, the taboo still makes sense. It makes far less sense today in urban areas where water for washing is plentiful, but the taboo is embedded so deeply in the culture that it still applies.

Perception as experienced by others can also be deduced by using many of the same sources and expertise that I recommended for deducing eliciting stories earlier in the paper. The web is a particularly valuable tool for this. Reliable sources of polling and survey data—such as Gallup's World Poll—can also be invaluable for this kind of research.

Perception, then, is in the eye of the beholder, and it is a function of both observation and recognition—or what we call mind-sets (sometimes also known as mental maps).

Mind-sets

Webster defines mind-set as a "fixed mental attitude formed by experience, education, prejudice, etc.," and thus it covers a range of cultural and cognitive considerations that lie behind perception. I would like to add that emotional or affective experiences also add to one's mind-set. From our perspective, a mind-set is the frame of mind we bring to bear on situations and events, based on our story or the relevant aspect of the story; and it provides a means for helping us to understand those events.

According to Segall, Campbell, and Herskovits, even anthropologists, who are trained to be as objective as possible, find themselves to be more ethnocentric and culturally conditioned than they had realized themselves to be.[17] In other words, their mind-sets played a larger role in their interpretation of observations than they had expected. Indeed, ethnocentrism has always been underestimated. Much of this section is derived from the book *Mindsets* by Glen Fisher, an experienced Foreign Service Officer, academician, and researcher.[18]

An example of ethnocentrism was given by the late Les Aspin while he was a Congressman. In an article in *Foreign Affairs,* he noted that U.S. intelligence estimates of Soviet military threats may have contained distortions because of the U.S. propensity to focus on hard evidence of technological capabilities rather than on Soviet intentions or on how they might use the technology.[19] Another closely related effect of ethnocentrism is attribution of motives. Fisher recounts two examples.[20] When he was a student, he went to rural Mexico as a volunteer helping with public health projects. These projects involved a lot of digging—

---

17. Ibid., 16.

18. Glen Fisher, *Mindsets* (Yarmouth ME: Intercultural Press, Inc, 1988).

19. Les Aspin, "Misreading Intelligence," *Foreign Policy* 43 (Summer 1981) 166-72.

drainage ditches, latrine pits, and such. The volunteer group tried to explain that they were there to "do good" and to help international understanding. The Mexicans couldn't believe those motives. They were convinced that the volunteers were there to dig for gold or oil. Another more serious example was the shooting down of the Korean Airlines Flight 007 near Sakhalin Island. The Soviets were blamed for being trigger-happy and disregarding human life. Later analysis suggested that the Soviet pilot who shot down the plane did not imagine that it could be anything but a foreign military plane conducting reconnaissance. He perceived according to his mind-set.

We all perceive according to our mind-sets. That means that we are predisposed to seeing things that we expect to see or want to see. Consider a trivial example, but one that makes the point about different perceptions. Imagine going to a brand new shopping mall with your spouse and a couple of teenagers. After two hours, have each person describe the mall and list the stores in it. It is almost certain that each will have a unique perspective, and the lists will be different. Yet each will have been exposed to exactly the same information. In other words, each individual is (unconsciously) very selective in what he or she observes. Our mind-sets can often cause us to add what is not there—as seen in the figure/ground examples from Gestalt psychology. Mind-sets can also cause us to subtract material that does not fit with our mind-sets—this is known as non-perception. There is a particularly dramatic example, quoted by Charles Tart, in a story from anthropological texts about Captain Cook.[21] When Cook sailed into a particular bay in the South Seas for the first time, the islanders gave not the slightest indication that they had seen his ship, even though it was right there in front of them. When a small boat left the ship to row ashore, the islanders saw it immediately and became alarmed. Apparently they thought the small boat had come out of nowhere because they had never seen a ship so large before, and therefore literally did not "see" it because of their mind-sets.

We need to realize that "facts" are not stand-alone, absolute truths. When looking at historical facts, we need to know how those facts and that history are remembered and by whom. Most people respond to images of facts that include what they think happened, their prejudices about it, and what was at stake. As Glen Fisher remarked, "international relations revolve around an interplay of images."[22] He also indicated that mind-sets may be as relevant in foreign affairs calculations as an inventory of development resources or a

---

20. Fisher, *Mindsets* 33-34.

21. Charles Tart, *Waking up: Overcoming the Obstacles to Human Potential* (Boston, MA: Shambala Publications, 1986) 104.

22. Fisher, *Mindsets*, 4.

point of international law,[23] and I would add that they are also as relevant as economic, technological, and military matters. Rather than trying to use some recent examples that have significant emotional content, to illustrate this, I will take two earlier historical examples, one from Joseph H. de Rivera's book, and another from Fisher. De Rivera discusses a cable that arrived at the State Department in early June, 1950. The cable stated that there was a significant arms build up along the 38th parallel in Korea and giving the North Koreans overwhelming superiority over the South.[24] This could have (should have) caused concern about a potential attack. Instead, it was perceived completely differently. The Ambassador to South Korea had been in Washington D.C. shortly before to request tanks and other equipment. The Assistant Secretary at the State Department and his staff thought that the cable was information supporting the Ambassador's request and did nothing about it. This is a good example of a narrow bureaucratic mind-set missing the larger picture.

Fisher quotes the example of the attack on the American Embassy in Teheran, and the holding of its diplomatic personnel hostage in 1979. The international response was that the Iranian action was irrational—it was a long way outside the limits of civilized diplomatic behavior, and we did not know how to deal with it diplomatically. Could we have anticipated this violent reaction to the Shah's admission to an American hospital? Possibly, if we had understood the depth of emotion among the Ayatollahs and their hatred of both the Shah and the United States, but we were not used to dealing with such emotional people. As Fisher describes, a tongue-in-cheek remark was made in the State Department, "You see, Jack, it is all a matter of the Gods. Now ours is a nice, reasonable, rational kind of a God. But the Ayatollah Khomeini's God—he's crazy!"[25]

I touched on ethnocentrism earlier. Real communication is two way. When we have dealings with others, it is important for us to understand our own mind-sets, biases, and prejudices just as much as it is for us to understand those of others. If we cannot do that, we are likely to suffer from the scourge of mirror-imaging and the problems that causes. Even in our own country and among fellow Americans, we have that problem. The stories of those in the Red States and those in the Blue States are quite different. Their attitudes about all manner of things, from immigration to the environment, can turn otherwise "normal" people into raging protesters and anti-protester protesters. Perceptions of events, issues, and policies depend on how they are

---

23. Ibid., 5.

24. Joseph H. de Rivera, *The Psychological Dimension of Foreign Policy* (Columbus, OH: Charles E. Merrill Publishing Company, 1968) 19-21.

25. Fisher, *Mindsets*, 7-9.

presented, and by whom, which brings us to the media and its role—a topic for discussion later in the paper.

Mind-sets are a function of culture, and vice versa—as perception over time can also influence culture. What is clear is that people in particular cultures, and especially sub-cultures, share similar mind-sets that simplify communication. One example of a small subgroup might be the husband and wife who can (and often do) complete each others' sentences. Another example might be a political group that sees conspiracies everywhere and vilifies people of another political persuasion. And then we have whole nations who regard major ethnic groups or other nations as unclean, ungodly, and adversarial. One of the interesting ideas here is that the more people are treated as if they are enemies, the more they become enemies. In other words, those who are perceived as enemies develop expectations that change their mind-sets, which, in turn, change their culture's orientation to others. In several discussions during the course of the writing of this monograph, various colleagues have remarked that in Iraq, while the behavior of young soldiers and Marines is entirely understandable, and while they are ordered to wear flak jackets and helmets and carry rifles at the ready, they appear to Iraqis to be treating all Iraqis, including women and children, as enemies. They contrasted this with the appearance of British troops who are ordered to take off their helmets and flak jackets once major combat is over. While concerns about force protection are certainly valid, it may be that offering a less threatening appearance might help to defuse the potential for confrontation. This would be an example of Robert Rosenthal's "expectation effect."[26] Although Rosenthal's experiments were conducted in elementary schools, to see whether teachers' expectations influenced their students' performance, and later the impact of clinicians' expectations on their patients' health, it is possible to extend the concept to any kind of expectation.

Understanding Mind-sets

When considering mind-sets, we may perceive them as barriers to our understanding: they are foreign; they don't make sense to us; we don't know the differences between the mind-sets of different tribes or groups living in the same area…and so on. Understanding them is difficult, and doing so may seem to be more trouble than it is worth. However, we need also to recognize the benefit of understanding mind-sets. This is a part of making the "invisible" visible. Once we understand them, then we can make use of that information for all the communications in which we wish to engage with a particular group or tribe. They will simplify our approach to IO, Psychological Operations (PSYOPS),

---

26. Robert Rosenthal, Biography. University of California, Riverside, June 6, 2007, at www.facultydirectory.ucr.

Strategic Communication (SC), and even Intelligence. What we need to look for, therefore, are patterns of thought—ideas that fit together as part of mind-sets—that are shared by cultures or groups.

Orientation towards the past or the future is one set of patterns. Fisher provides the example of Argentina trying to take back the Malvinas (Falkland Islands) by force in 1982.[27] The British were taken by surprise, since they had gained sovereignty over the islands early in the 18th century, and they were bemused by Argentina's position. But to the Argentinian mind, the conquest was still vivid and alive; they were convinced that their's was a reasonable posture, and they expected world support for it. We have seen similar historic events raised during the conflicts in the Balkans and throughout the various conflicts and wars in the Middle East. This is not a mind-set that is shared by the United States and Britain and some of Western Europe. We tend to think of the past as being past, and our orientation is to the future.

Orientation towards some set of beliefs also provides a mind-set. For instance, we have recently seen the problems that have arisen in academe by adherence to political correctness, as demonstrated recently at Duke University. The eighty-eight professors who took out an advertisement supporting those who were denouncing and threatening the lacrosse players tended to be from those departments that promote political correctness. It is likely that they were so immersed in their own story—myths of white "jocks"—that they couldn't conceive that the students were innocent.

Another example, in which Fisher provides both his own observations as well as quoting from Glen Dealy, illustrates some of the key differences between the United States and Latin America.[28] American society is positive, optimistic, and "can-do." It places great emphasis on individual achievement and the rewards to be gained from it, and it emphasizes efficiency and progress by the management of "things." It is essentially middle class, and has middle-class values, and the real source of power is economic power. People who are highly regarded in society (even including movie stars and sports personalities) tend to have these characteristics, although working hard for a living provides a greater sense of success. In Latin America, class plays a much greater role. Being who you are counts for more than what you have achieved. Indeed, work is to be avoided wherever possible. Education focuses on the development of personality and social skills, and the emphasis is on managing relationships rather than on managing things. Latin Americans are more fatalistic, as seen in their arts, and even a sense of individualism is defined more in terms of

---

27. Fisher, *Mindsets*, 41.

28. Ibid., 52-55, 64-65, 110-111.

personal dignity than in terms of achievement. Their leaders are chosen for their forceful personalities and eloquence and for their appearance of being able to take charge—in other words their social power. Together, these kinds of patterns in mind-sets can provide a useful background from which to analyze communications and also to direct communications in a more meaningful and influential way.

We need to realize that these mind-sets and group patterns affect everyone from a culture, even those who may be involved in diplomacy, the Armed Forces, or multinational corporations and those who have lived abroad for many years. When they make decisions, set a policy, or execute one, they reinforce their views about the rightness of that policy or action while trying, at the same time, to pay attention to the culture in which they are operating. When the policy or action is taken in a country remote from the country of the actor's origin, then it is likely to be less acceptable or viable to that country, but it may "sell" better (to Congress, constituents etc.) in the country from which the actor comes.

Fisher suggested that one of the best ways to understand mind-sets is to observe ordinary behavior, such as noting how praise and censure is given, how and what children are taught, and what makes people sentimental or arouses emotional reactions.[29] Insight can be found in humor, cartoons, literature, poetry, and especially in children's stories. These may not seem likely sources of intelligence for military organizations, but they should be if we are interested in understanding and influencing various countries, cultures, or groups. In addition, as mentioned earlier, the internet, which was not around when Fisher wrote his book, is also a good source of information about mind-sets, especially those of specific activist/terrorist target groups.

One of the strongest influences on mind-sets is the Fourth Estate. This is deserving of a section all on its own.

The Media

We cannot overstate the importance of the media in developing the stories, the mind-sets, and the perceptions of those in our own country—policy makers and the general public—as well as those in our adversaries' countries. Moreover, journalists are rarely the objective reporters that they would like people to believe; rather than being simply the providers of information about events, they have increasingly become part of the events. Their biases have been recorded not just by people of different political persuasions, but by members (or former members) of their own fraternities,

---

29. Ibid., 55-56

such as Bernard Goldberg from CBS, who has focused on the media's biases, and John Stossel from ABC, who has focused on its lack of understanding of science and economics.[30] Good news has never sold as many copies of newspapers or improved the ratings of television news stations as bad news, so bad news is more widely reported. However, the repetition of horrific images and negative stories must surely have an impact on the mind-sets of frequent viewers. In recent years the media have become more political and, in many respects, anti-American, giving predominant voice to those with grievances and political axes to grind. American news media are not the only ones with problems. Recently, as reported in the New York Times on May 3, 2007, Howayda Taha from Al Jazeera was convicted in Egypt for purportedly fabricating a documentary that accused Egyptian police of torturing their prisoners.

The media are so dominant in today's world that diplomacy, international affairs, and military engagements have to take note of the media's watchful eye and conduct their business in that arena rather than at the negotiating tables and conference rooms of yesteryear. It places enormous constraints on what can be done and how it can be accomplished when leaks or tactical errors can be viewed around the world within minutes and can lead to huge strategic problems. Recently, Tom Ricks, writing in the Washington Post on July 1, quotes from Captain William Ault's argument in a recent issue of *Armor* magazine. How could "the undisputed military power in the world" be losing a war against lightly armed insurgents? Ault's answer: "The Media did it." Ricks goes on to say that the media assists insurgent forces by continually maintaining pressure on the supporting government and military establishment. The same week, Oliver North, expressed his view of the power of the media and their tendency to exaggerate any statement or event to further their point of view. Commenting on the media's coverage of Richard Lugar's recent statement about the war in Iraq losing "contact with our national security interests in the Middle East and beyond," North noted "America's media elites....have determined the outcome of the war against radical Islam will be decided not on the battlefields of Iraq, but in the corridors of power in Washington. And about that, they may very well be right."[31]

---

30. Bernard Goldberg, *Bias: a CBS Insider Exposes how the Media Distort the News* (Washington DC, Regnery Publishing, Inc., 2002); John Stossel, *Myths, Lies, and Downright Stupidity* (New York, Hyperion, 2006).

31. The text of North's article may be found online on the Human Events website, at http://www.humanevents.com/article.php?id=21351. Posted 29 June 2007; accessed 5 September 2007.

Another problem is that of the "specialist." Most of the situations in the world today are complex, and a real understanding is likely to require a multidisciplinary approach, involving anthropologists, psychologists, regional specialists, military analysts, and many others. All of that takes time, and so the media rely on experts or specialists who come from a single discipline, or on retired military officers, some of whom grasp the implications of this new form of warfare, and some of whom do not. Who is the public to believe? From whom do they get their stories and mind-sets? Generally, from those whose values and politics they share. Thus, any serious analysis of mind-sets needs to include the role of the media and the mind-sets of its prime commentators.

Group Identity

Everyone is a member of many groups—most of which are taken for granted—national, race, ethnic, gender, interests, sports, school/university, and many other organizations. We see the strength of identity among serious fans when one football or baseball team is beaten by another. In politics, we see candidates identifying with their constituents, rolling up their sleeves, speaking with slightly different accents, and using different colloquialisms in order to demonstrate their solidarity with and commitment to them. Clearly, the groups in which we are interested are the terrorist and insurgent organizations currently operating across the Middle East, the Far East, and with cells in Europe and the United States. These tend to be Islamic (both Sunni and Shia), Arab, Indonesian, and Philippine. While we may also be interested in drug and terrorist groups in Latin America and Africa and the potential for conflicts with China and North Korea, let us stay with the more obvious terrorist groups for the moment.

These Islamic terrorist groups are formed around religious ideals and ideologies. What are the essential qualities of these groups? Kenneth Hoover discusses some of the key attributes of identity, including competence and integrity.[32] In this definition, competence means the ability to have productive social and personal relationships that can be validated by the group. Integrity includes beliefs—religious, ideological, national/tribal—and it involves transactions between the individual and the group that involve loyalty and ethnic ties. It also includes shared meanings that the individuals create or acquire. This suggests that values and motivations are key dimensions of group identity.

---

32. Kenneth Hoover, *The Power of Identity: Politics in a New Key* (NJ, Chatham House Publishers, 1997) 19, 49-52.

Together with a number of colleagues, I have researched, designed, and developed models of values and motivations that have been used in the industrialized world since the early 1970s. Those models have been used by commercial organizations for strategic planning, marketing, advertising, new business and product development, R&D planning, and communications—all areas that are also relevant for IO, PSYOPS, SC, and Intelligence. One of these models will be described briefly in the section on Cultural-Cognitive Dimensions and in greater detail in the Appendix.

Mind-sets of the Arab/Islamic World

Since September 11, 2001, we have focused much of our military activities on Iraq and Afghanistan and on al Qaeda and other radical Islamic groups from Europe to Indonesia and the Philippines. The cultures, stories, mind-sets, and perceptions of Arabs and other Muslims are so very different from ours that it is worth mentioning a few key concepts here, although this monograph is not about Radical Islam.

With one or two exceptions, most Islamic countries have an authoritarian epistemology based on the Qur'an. This means that their understanding of everything—science, economics, justice, medicine—is based on the Qur'an. Some sects are more strict than others, with the Wahhabi madrassas in Pakistan—where the Taliban are educated—teaching a particularly strict and radical form of Islam. In this kind of education, there is no concept of the individual learning anything, except by rote. He has to be able to memorize and interpret the Qur'an to obtain answers to any questions. Creativity, intuition, and initiative are virtually forbidden. Conversations with an individual who was trying to run an electronics company in an Arab state several years ago suggested that, even when administering examinations to ensure that the electronics engineers were competent, the questions had to be identical to ones the new engineers had already seen. Officials from the government objected to different questions, even when they were similar to those in the texts. This discouragement of individual initiative and insistence on the authority of the Qur'an is a likely cause of the poverty and backwardness of the Arab countries. Even when the ruling families and other individuals have been educated in the West, they do not have sufficient power or, more likely, the will to change things.

These kinds of circumstances, together with the authoritarian epistemology, have prevented many Islamic countries from developing the kinds of economies that they see in the West. David Pryce-Jones goes so far as to say "Modernizing in this respect, it seems, fatally wounds the core (Islamic) identity."[33] And he then suggests that the failure to develop economically and politically generates self-pity, especially when they have experienced so many wars, civil wars, coups, and other events that have caused suffering and poverty for so many Islamic populations.

In conversations with a Pakistani friend, he commented that young people in Pakistan today don't know who they are. He said that many of those from poor families, and who live in Pakistan, seem caught between two worlds—the old, orthodox world of their parents and tribes, and the new world full of modern technology and consumer goods. They see the material benefits of the West, yet without the education and understanding of what has created such wealth, they don't know what to do about it, or how to attain it. There are exceptions, of course, especially those from wealthy families who have been educated in the West and who may live in the West. But he said that even among those young people, there is a great deal of dissatisfaction because they cannot find meaning in their lives. Many of their wealthy parents are not very religious, and their children are becoming more orthodox, partly as a rebellion and partly because they are seeking meaning that they cannot find elsewhere. This reminded me of the interviews that McConnell conducted with young extremists, many of whom expressed similar sentiments.[34] It may well be that the "home grown" terrorists and insurgents from Western countries cannot find meaning in their lives; their families have become more secular and less concerned about their religion, and even the wealthy middle class may feel a sense of being second-class citizens. The opportunity to be heroes, to have meaning—even if it means killing others and dying for the cause—could have an appeal for these young men (and perhaps women).

Several questions need to be raised here:

- What causes teenagers or young men, who have previously been quiet, serious students or members of their communities in various countries in both the West and the rest of the Arab world to go to Pakistan and train in their madrassas, and then to fight in Iraq, Afghanistan and elsewhere?
- What are the tipping points?"
- What could be done to prevent them from going?"

---

33. David Pryce-Jones, *The Closed Circle: an Interpretation of the Arabs*, (Chicago, Ivan R. Dee, new edition 2002), xi.

34. McConnell, *Stepping Over,* passim.

Even Osama bin Laden, who came from a wealthy and privileged family, exhibited authentic self pity, as Pryce-Jones noted, in his statement after September 11, when he said "Thanks be to God that what America is tasting now is only a copy of what we have tasted. Our Islamic nation has been tasting the same for more than eighty years of humiliation and disgrace, its sons killed and their blood spilled, its sanctities desecrated." [35]

Jabal al-Din-al-Afghani (1838-1897) was probably more influential than any other Muslim in shaping the ideas of the Muslim world. As Patai remarked "Afghani recognized and stated emphatically that the British and French conquests in the Middle East…were made possible by science and that, therefore, the Arabs must acquire science if they want to liberate themselves from Western domination."[36] He was amazed that the Christians had invented the cannon rather than the Muslims.[37] Afghani continually applied the term "backwardness" to Muslims, setting up the tension of self-pity and failure resulting from adherence to Islam, which he exacerbated by exhorting Muslims to follow Islam as he thought it should be. All of this set the scene for the Muslim world, which had been well on its way to conquering much of Europe a few centuries earlier, to succumb to feelings of shame.

It is interesting to note that, in the West, scientific investigations began during the Renaissance (the rebirth of the Classical era) and became formalized during the Enlightenment. This was the time of a major paradigm shift for the West, away from the authoritarian epistemology of medieval religious doctrine and towards an empirical epistemology based on the scientific method. The Islamic world has not been through a similar shift, which could explain its current predicament.

Patai discusses at length the problem of Arab stagnation and loss of initiative between the 7th and 13th centuries, and he quotes many eminent Arab scholars who have suggested that Arabs were living in the Middle Ages until Napoleon arrived in Egypt.[38] In the three hundred years since then, many scholars have suggested different paths to the future, including the separation of religion and science. However, in the ensuing self-analysis, since they could not accept that the values of the infidel were better than the values of the faithful, Muslims came to the conclusion that they had distorted their own values and had become different from what Islam had really taught. This was

---

35. Ibid., xii.

36. Raphael Patai, *The Arab Mind* (New York, Hatherleigh Press, Revised edition, 2002), 291.

37. Pryce-Jones, *Closed Circle*, 87.

38. Patai, *Arab Mind*, 261-283.

the beginning of the religion's revival, the attempt to rediscover the old values and purify Islam.

In Patai's conclusions on the Arab mind, he discusses various areas, including language: "It is in this specific, emotionally colored quality language has for the Arab, in the sensate satisfaction he derives from the sound, rhythm and cadence of Arabic, that one must seek the psychological bases of his inclination to rhetoricism, exaggeration, overassertion and repetition, and of his tendency to substitute words for actions."[39] The psychology of Arabs is to be found in the two traditional components of society—the pre-Islamic Bedouins and then the Islamic component. The pre-Islamic Bedouin element contains the sense of kinship, loyalty, bravery, manliness, aversion to physical work, and a great emphasis on honor, "face," and self respect. It also includes raiding, blood revenge, and hospitality. The sexual honor of women is especially important, and the honor of the woman's entire paternal family depends on it. The Islamic element is seen in the way religion permeates every aspect of life, providing a sustaining force and direction. Predestination or fatalism is also a hallmark of the Islamic element; everything is attributed to the will of Allah. Generally the Arab temperament is one of conforming, but the culture does provide a means through which suppressed emotions can break out. These are periodic flares of temper, anger, aggression, and violence that are condoned by society. Patai believes that there is a lack of correlation between thoughts, words, and actions, with the Arab thoughts and words being more idealistic and independent of reality than action. This is reflected in Fouad Ajami's book about the Arab intellectuals in the post World War I period, who longed for the Arab world to embrace modernity. The chapter on the suicide of Khalil Hawi is a particularly poignant account.[40] It describes the personal anguish of a Christian Arab from Lebanon who was angry at what was happening to his country, who saw what was coming yet felt powerless to do anything about it.

One key concept that Patai examines at length in his book is that of the Arab quest for unity.[41] Arabs have a deep conviction that, despite their being citizens of different states and members of many social and ethnic groups, there is one Arab nation, and that all Arab countries are part of a single homeland. This was the supposed raison d'etre of the Socialist Arab Ba'ath Party, although in practice it has done the opposite. Nonetheless, the restoration of the Caliphate is a highly motivating vision.

---

39. Ibid., 326-327.

40. Fouad Ajami, *The Dream Palace of the Arabs* (New York, Pantheon Books, 1998), 26-110.

41. Patai, *Arab Mind*, 359-364.

Since Patai wrote and then revised his book, many events have occurred. Yet today's terrorists and insurgents seem to represent an uneasy truce between traditional Arab (Bedouin) and Islamic mind-sets and modernity in the form of science. While the radicals permit no creativity in general, they exhibit great creativity in terms of tactics and the development of IEDs, bombs, and other weaponry. They are innovative in their uses of technology such as cellphones. But behind them are still the concepts of revenge, honor, and "face" mixed with resentment and envy of the West.

So What Can DoD Do?

First, it seems to me that what is needed in the short term is a way of developing an understanding of cultures, stories and mind-sets that can give us a much greater awareness of what any adversaries anywhere are thinking and doing and what they are likely to do next, even if it gives only an 80% solution. The Department of Defense (DoD) needs that edge now in its counter-insurgency and counter-terrorism efforts. The approach the Applied Futures' Team is taking in order to accomplish this is the focus of this monograph—we call it Cultural-Cognitive Dimensions Analysis.

Second, for the longer term, it seems imperative that The United States conduct a great deal more research on cultures, stories, and mind-sets than it has heretofore. This research should be multidisciplinary and should go far beyond anything that is being done currently by academe or government. And the results should be provided to key agencies from the National Security Council to the State Department, DoD, and the Intelligence Community. Much current research is being done through single disciplines, on single countries, cultures, or regions of interest, without any attempt to identify patterns and similarities that could provide cross-cultural insights

In terms of the research needed, as mentioned earlier, Fisher suggested that one of the best ways to understand mind-sets is to observe ordinary behavior, either directly or as reflected in humor, cartoons, literature, poetry, and in children's stories. In addition, the internet is also an excellent source of information about mind-sets, especially those of specific activist/terrorist target groups. Radio Free Europe/Radio Liberty has just (25 June) released a report—*The War of Images and Ideas*—on the breadth and sophistication of Iraqi insurgent media.[42] It focuses on Sunni insurgents, as the Shia have access to established media outlets. We must listen to the conversations of our target audiences—on the web as well as in coffee shops and market places. We must

---

42. Daniel Kimmage and Kathleen Ridolfo, *Iraqi Insurgent Media: the War of Images and Ideas*. Available on the Radio Free Europe/Radio Liberty website at: realaudio.rferl.org/online/ OLPDFfiles/insurgent.pdf. Accessed 7 Sep. 2007.

watch their videos, their movies, and their television shows, and we must learn what is so funny about their jokes and cartoons, which often contain real truths about how people are feeling. We can make use of surveys and polls such as the Gallup World Poll. We need to conduct focus groups, facilitated by locals familiar with the particular dialects and cultures, to help evaluate data from these sources and to reveal stories and mind-sets. To this end, DoD must recruit and develop interpreters and translators who are current in the languages and in modern slang and colloquialisms, experts who understand the pertinent history, including recent events, and specialists in semiotics. These sources of information and the approaches used to identify and clarify stories can also be used to understand perceptions and mind-sets.

All these forms of research can be used both for longer-term research and to flesh out and test Cultural-Cognitive Dimensions Analysis as outlined in the first recommendation.

Cultural-Cognitive Dimensions Analysis

Cultural-Cognitive Dimensions Analysis is a method to enhance our understanding of what we are seeing our adversaries do and what meaning and motivation we attribute it to. It is based on an approach for seeing through our adversaries' eyes (mind-sets), and deducing how they understand the situation, what they are thinking about it, and what they are thinking of doing about it. This will provide a greater understanding of truth for them as well as us and of our perceptions of them and their perceptions of us; this in turn will provide us with better insight into the possible and probable consequences of any actions that we take in the light of those understandings.

The author has identified a limited number of key cultural and cognitive dimensions that can be used to model any culture or country, and which the members of our team believe are critical for military/political purposes. We are not attempting to be comprehensive here; we are looking for 80% solutions, not perfection. Nevertheless, the list is fairly comprehensive:

- Epistemologies
- Ways of Thinking
- Values, Beliefs, and Motivations
- Approaches to Life
- Approaches to Understanding
- Approaches to Power
- Measure of Achievement
- Religious Beliefs

- Concern about Honor
- Concern about Shame
- Strategic Time
- Tactical Time
- Group Orientation
- Assertiveness
- Attitude towards Death
- Reactions to the Foreign

Each of these dimensions is described in more detail below. These are not necessarily orthogonal, indeed some are subsets of others, yet we have called them out separately because of their importance. One of the areas that we shall be researching is to identify the relationships between the dimensions, so that we can develop templates or models for each country or culture.

I have developed this list of Cultural-Cognitive Dimensions to enable us to describe cultures according to several key characteristics. They have been further refined based on extensive research into cultures and cultural modeling from anthropological, psychological, and sociological perspectives. The sources for this research are included in the references.

Epistemologies (authority-based – *to* – empirical)

Starting in the Renaissance, we in the West have adopted an empirical, scientific approach to how we know things. We use the scientific method— positivism, objectivism, and reductionism. We analyze things, break them into component parts, and reassemble them. In some non-Western countries, the way of knowing things is much more authority-based (as was the Western approach during the Middle Ages.) In the Arab Middle East, for example, the approach is authoritarian, which is why the Mullahs have so much influence. The Qur'an is the prime source of knowledge. People use it, and are required to use the Mullahs' interpretations of it, to make sense of their situations. David Cook has written about the defeat of the Taliban and how their best clerics are now poring through the Qur'an to try to find out how Mohammed overcame his defeats in battle and went on to victory, so that they might emulate him.[43] To us this kind of "knowing" is superstition; thus, we do not even know how to go about understanding such people. If we do not share epistemologies,

---

43. David Cook, "The recovery of radical Islam in the wake of the defeat of the Taliban", in *Terrorism and Political Violence*, 15, 1, Spring 2003, 31-56.

then this is the first dimension that we need to consider when dealing with another culture.[44]

Ways of Thinking (linear – *to* – holistic)

Research on cultural differences in Ways of Thinking has been conducted by Richard Nisbett.[45] This dimension has great importance to us, since we in the West (and particularly the United States) have a very linear, rational (Cartesian) approach to thinking. We pride ourselves on our analytical capability and our ability to separate out logic from emotion. In doing so, however, we often ignore contexts and the interdependencies that are of critical importance to other cultures. Indeed we find it difficult to imagine how people who think holistically operate. Yet, as mentioned earlier, Nisbett has discovered that cultures in the Far East tend to have a much more holistic way of thinking. Cultures in the Middle East seem to be somewhere between the two. We would consider this to be the next dimension that we should understand when communicating with other cultures.[46]

Values, Beliefs, and Motivations (sustenance driven – *to* – self actualization)

This dimension is based on Maslow's theory of motivation and hierarchy of needs.[47] Since the early 1970s, research has supported the idea that people have a set of values, beliefs, and motivations that are relatively consistent, that are culturally based, and that underpin everything they do. These values and beliefs manifest through time as attitudes and lifestyles, and in the short-term as behavior and perception.[48] On one side of the continuum are cultures that are sustenance driven. They are concerned with meeting their basic needs and, even when they have the things they need for survival, they tend to focus on holding on to what they have, including tradition.

---

44. See Bernard Lewis, *Islam and the West*, (Oxford, Oxford University Press, 1993) and Edward W. Said, Orientalism, (New York, Random House, 1978) for two contrasting but useful perspectives on the clash between Western and Middle Eastern epistemologies; see also W. Kirk MacNulty, "The paradigm perspective", *Futures Research Quarterly*, 5, 3, 1989 35-54.

45. Nisbett, *Geography of Thought*, passim.

46. See also Helen A. Klein, "Cognition in natural settings: the cultural lens model" in Kaplan (Ed) *Cultural ergonomics, advances in human performance and cognitive engineering* (Oxford, Elsevier Press, 2004), and K. Peng and R Nisbett, "Culture, dialectics and reasoning about contradiction", *American Psychologist*, 54, 741-754.

47. Abraham Maslow, *Motivation and Personality*, (New York, Harper Row, 1954).

48. Christine MacNulty and Leslie Higgins, *Applied Futures Social Change Program Reports* (London, 1988-1993).

In contrast, self-actualized cultures (and people) are not focused on basic survival needs. They are driven by intrinsic goals. They are psychologically mature and place more importance on personal accomplishment (such as a career, or to relationships) and on people than they do on material things. Values of this sort represent the third dimension in order of importance when we do not share epistemologies and ways of thinking, and the first in order of importance when we do. They also play a large role in group identity.[49] See the Appendix for more detail.

Approaches to Life (being – *to* – doing)

This may seem like a peculiar notion to Americans, who are inveterate do-ers. But there are parts of the world, both Middle and Far East, where the society has a much greater acceptance of what is, and very little inclination to change it. In the Middle East, the phrase *"If it be the will of Allah"* is used constantly, and events and circumstances are just accepted without the kind of questioning that Americans would indulge in. In the Far East, Buddhism encourages a similar, almost passive acceptance of life. Americans do not understand these attitudes. As a result we too often try to force issues far too fast for the people of those regions. An illustration of this is the American tendency to try to avoid or speed up the process of sitting down with Middle Eastern or Asian counterparts to drink tea. To Americans the process is a waste of time; they should be getting on with business. To the others, it is a process of becoming comfortable with the Americans—as they do with all guests—and not conducting business until they feel in harmony.[50]

Approaches to Understanding (thinking – to – feeling)

In the West, with our linear approach to thinking and our empirical epistemology, we have an intellectual approach to understanding. We expect to be able to analyze things intellectually, and we pride ourselves on being objective. This contrasts with Middle and Far Eastern cultures that place a much higher value on feelings and emotions. Clearly this dimension is related to the others mentioned here, but we have called it out separately because of its importance in developing and understanding communications and reactions.

---

49. See SRI International, *VALS Program Reports*, (Menlo Park, CA, 1978-1981); Ronald Inglehart, *Culture Shift in Advanced Industrial Society*, (Princeton, NJ, Princeton University Press, 1990; and Ronald Inglehart, *Modernization and Postmodernization*, (Princeton, NJ, Princeton, University Press, 1997).

50. Martha Maznevski et al "Cultural dimensions at the individual level of analysis," *International Journal of Cross-Cultural Management*, 2002, 2.

Approaches to Power (centralization – *to* – decentralization)

The difference between centralization and decentralization is fairly obvious, and its importance lies in understanding how tightly or loosely the government and other major institutions control the people. This tells us how much freedom the citizens have and, therefore, how well, or little, we as Americans may be able to influence them as compared to the influence of their own government. It also indicates the press and media's degree of autonomy with regard to reporting events.[51]

Measure of Achievement (material – *to* – social)

This is not so much a dimension along which a country can be measured, as it is part of the success mind-set. Earlier I mentioned that achievement is a strong American value, and it is usually measured in terms of material success—acquisition of material wealth and power. This contrasts with the Latin American success, which is measured in terms of social relationships and social status.[52] It appears that within terrorist organizations success is based on numbers killed, amount of daily life disrupted, and probably minutes/hours of broadcast time and inches of newspaper reports on their activities.

Religious Beliefs (critical for all aspects of life – *to* – lack of religious beliefs)

Islam is more than a religion. It is a way of life that permeates everything in the Arab world, including its governments and system of justice. *Shari'a* means Divine Law. While there are interpretations of the Qur'an that are far less fundamentalist than we see in many Middle Eastern countries, those interpretations most influential at present are the fundamentalist ones. These have strict rules about women, dress codes, behaviors, and so forth. While many of the young people envy our freedom, even they regard us as far too materialistic and licentious. They do not see our approaches to Christianity or Judaism as being religious in their sense, even if they recognize those religions. Since they do not hold us in high regard, it makes our work of influencing them very difficult.[53]

---

51. See the related ideas of the Power Distance Index in Geert Hofstede, *Cultures and Organizations*, (London, McGraw Hill, 1991), and Hierarchy in Shalom H. Schwartz, "Beyond individualism/collectivism: new cultural dimensions of values" in Kim et al (Eds) *Individualism and Collectivism: Theory, Methods and Applications*, (Newbury Park, CA, Sage, 1994) 85-119; and in Mazneski, "Cultural Dimensions."

52. Fisher, *Mindsets*.

53. For the following discussion of Religious Beliefs, Concern about Honor, and Concern about Shame, see Patai, *Arab Mind*; Pryce-Jones, *Closed Circle*; and Lewis, *Islam and the West*.

Concern about Honor (low – to – high)

This and the next dimension are related to Values and Motivations, but I believe that they are important enough to be called out separately. Honor is of great importance to everyone in the Middle and Far East, and it is a critical value for individuals, families, tribes, and nations. In contrast, outside the Armed Forces, in the West, the concept of honor has all but disappeared. Thus we do not pay anywhere near enough attention to this value in our dealings with people of other cultures.

Concern about Shame (low – to – high)

This dimension is the opposite of honor, but because it is so critical in its own respect, I consider it to be a key dimension. The Arab Middle East, especially, is very concerned with "saving face" and not bringing shame to oneself or one's family. (this is true of Far Eastern countries too). In the West, the concept of shame has almost disappeared. According to many of its own commentators, the Arab Middle East feels lacking in self respect because of the importance people place on how they are perceived, especially by the West. This often leads them to resent any kind of aid that we give them, and to resent us in turn. They are typically quick to see condescension in any communication or aid that is offered.

Strategic Time (short – to – long)

This dimension is about a culture's sense of history and the understanding brought about by that sense of history. It permeates every story, every perception, and every decision. It is also about the time over which they expect their actions to play out. In the United States, we have probably the shortest strategic time of anywhere on the planet, and this can be seen in our desire for everything to happen "right now." This can be a real handicap for IO and PSYOP, which can take considerable time before results are seen. China likely has the longest strategic time. The Middle East is generally somewhere in the middle: radical Islam talks about the re-conquest of Europe and the re-establishment of the Caliphate on the one hand, while countries such as Iraq have a very short history of their various tribes and religious groups working as a nation.[54]

Tactical Time (short – to – long)

The concept of tactical time is related to that of Strategic Time, but it is focused on the time taken to respond to events. A trivial example of

---

54. See Hofstede's Long Term Orientation in, Cultures and organizations.. Also related to this dimension is the concept of Time Horizon: see, Klein "Cognition in natural settings."

long Tactical Time is that of Finland where, at the end of a presentation, the audience expects to wait for several minutes before asking questions. They want to ensure that they have the time to think about what has been said, so that their questions are well considered. They regard it as rude to ask a question too quickly, unlike Americans who are ready to ask questions before the speaker has finished. The delay in the Radical Islam response to the Danish Cartoons raises a question: What caused the delay, and could we expect similar delays in responses to other events? This is a new concept that we are working on, and one which does not seem to have been explored by others. It could have significant implications for IO and PSYOPS.

Group Orientation (individualistic – to – collective)

This is a measure of the degree to which the society regards itself as a single entity or as a collection of individuals who are simply members of that society. On the individualist side we find societies in which the ties between individuals are loose: everyone is expected to look after him/herself and his/her immediate family. On the collectivist side, we find societies in which people from birth onwards are integrated into strong, cohesive in-groups: often extended families with uncles, aunts, and grandparents who continue protecting them in exchange for unquestioning loyalty. The word 'collectivism' in this sense has no political meaning. It refers to kinship in the group, not to the state.[55]

Assertiveness (masculinity – to – maternalism)

Although this dimension began by referring to the distribution of roles between the genders, Hofstede's work indicated that women's values differ less among societies than men's values.[56] Moreover, men's values from one country to another contain a dimension from very assertive and competitive to more caring and concerned values. In work conducted by Applied Futures, masculinity has come to mean an outgoing competitiveness and aggressiveness, while maternalism implies more caring and concern, even passivity.[57] Japan is the most Masculine country yet examined, while Sweden is the most Maternal. The Middle Eastern countries tend to be more Maternal because of their passivity and fatalism.

---

55. See Hofstede, *Cultures and Organizations*, and Maznevski et al, "Cultural dimensions," and the concepts of Affective and Intellectual Autonomy in Schwartz, "Beyond individualism/collectivism."

56. Hofstede, *Cultures and Organizations*.

57. MacNulty and Higgins, *Applied Futures Social Change Program Reports*.

Attitudes Towards Death (acceptance/avoidance)

This is not a dimension in the sense that we can measure countries along it as we can with the other dimensions; rather it is about the impact of attitudes about death on behavior. For instance, the Islamic suicide bombers do not appear to be afraid of death—they go to it willingly—and they are regarded as martyrs. There may be various reasons for this. Not only do they get the promise of heaven and virgins, they may remove the shame of poverty and unemployment and bring honor to their families. In addition, since some get paid for doing it, they also bring financial rewards to their families. While Christians also expect to go to heaven after death, they do not seem to welcome death in the same way. Indeed, they do everything they can to avoid death. This makes them much more vulnerable to acts of terrorism than almost any other culture. Although we are probably not going to be too concerned with studying Buddhists, they tend to have a calm acceptance of death, perhaps even welcoming it.

Reactions to the Foreign (open – to – closed)

While this dimension has some relationship to the political system, it is also cultural and related to values. Some countries are open to foreigners and foreign ideas (e.g., the USA, UK, and Canada). Some are open to innovations and assimilate them easily, but are not quite so open to foreigners (e.g., Japan). Other countries are quite closed to new ideas (e.g., Burma). This dimension is likely to be useful in determining how information operations might be conducted.[58]

In figure 2, I illustrate educated guesses about how the U.S. and Iraqi cultures would fall on some of these dimensions. I have also included an ally—France—to indicate that, even in the West, there are differences, which makes it necessary to understand allies from these same perspectives.

Where the countries' positions are close together, it means that they have greater likelihood of understanding one another—although even here, we need to be aware of the totality of the dimensions and how they manifest in each country. Where they are far apart, it means that they are likely to have real differences in perception, understanding, motivation, and behavior. At present, the positions of different cultures on these dimensions will have to be estimated using the subjective judgment of experts. Later, our team hopes to be able to quantify them.

---

58 Ibid., For a description of Western versus Arab openness, see Patai, *Arab Mind,* and Lewis, *Islam and the West.*

FIGURE 2. CULTURAL DIMENSIONS – EXAMPLES

For instance, as an example of the differences between us and an Islamic culture, let us imagine communications written in our normal fashion—applying logic in our usual linear style—and the content of our message is based on some empirical or scientific analysis. How will the Islamic culture (authoritarian, and with a more holistic way of thinking) view our communication? In May of 2006, President Ahmadinejad of Iran sent a personal letter to President Bush. According to media reports, the letter was about sixteen pages long, rambled on in some unintelligible way, contained many quotations from the Qur'an, and appeared to be asking President Bush to convert to Islam. The media's response was that the letter was the ramblings of a madman. I have not seen a translation of the letter, but it would not surprise me if it was a fairly typical Islamic/Iranian communication (from an authoritarian epistemology, and holistic way of thinking). President Ahmadinejad was not educated in the West, and therefore probably lacked the knowledge of Western diplomatic language. The letter certainly contained the first phase of a typical 3-phase Islamic threat—conversion, then coercion, then invasion. But because we (or our media) were viewing the letter through our Western mind-set, and probably through stories that we have told ourselves about Ahmadinejad, we could not give it credibility.

When we, in the United States, engage in IO, PSYOPS, or SC, we do employ foreign nationals to help us craft messages. However, we need more than educated people who can translate what we are wanting to say, we need to ensure that we are addressing the appropriate mind-sets in the appropriate way. When we engage in Intelligence operations, we need to ensure that we are understanding what is really being communicated, not just what a translator or interpreter tells us.

The Future of Cultural Understanding

Each of these Cultural-Cognitive Dimensions needs to be developed and fleshed out further for each culture under consideration. I believe that the material in this monograph will provide a good starting point for anyone who is working to understand adversaries and to develop means for dealing with them. The Applied Futures Team (composed of Applied Futures, Inc., Cognitive Performance Group LLC, and Alidade Incorporated) is working to identify the relationships between these dimensions in order to develop templates and models for applying them to any culture rapidly.

But there are two further elements to the whole picture.

Cognitive Task Analysis (Cognitive Performance Group), which provides a perspective on culture and decision mechanisms in order to identify the appropriate "lens" through which to assess a society's behavior, and thereby understand how to influence it. This analysis includes Sense-Making and Decision-Making at group and individual levels.

Complex Network Analysis and Complex Systems Research (Alidade), which provides an understanding of how collective behavior propagates and amplifies through the structure, dynamics, and evolution of networks (such as media, social relationships or other networked cultural [political, military, economic, social, infrastructure, information – PMESII] artifacts).[59] This includes identification of Tipping Points.

Together with the Cultural-Cognitive Dimensions, these two areas form a technique our team has called Cultural-Cognitive Systems Analysis (CCSA)[SM] This provides the most comprehensive approach for planning and assessing IO, PSYOPS, Effects Based Operations (EBO), and SC that exists.[60] A diagram of that approach, developed together with the Cognitive Performance Group and Alidade Incorporated, is shown in figure 3. But that is another story...

59. The acronym PMESII stands for Political, Military, Economic, Social, Infrastructure, Information systems.

60. Christine MacNulty, William Ross, and Jeffrey Cares, "Cultural-Cognitive Systems Analysis (CCSA)[SM]" – The subject of many presentations.

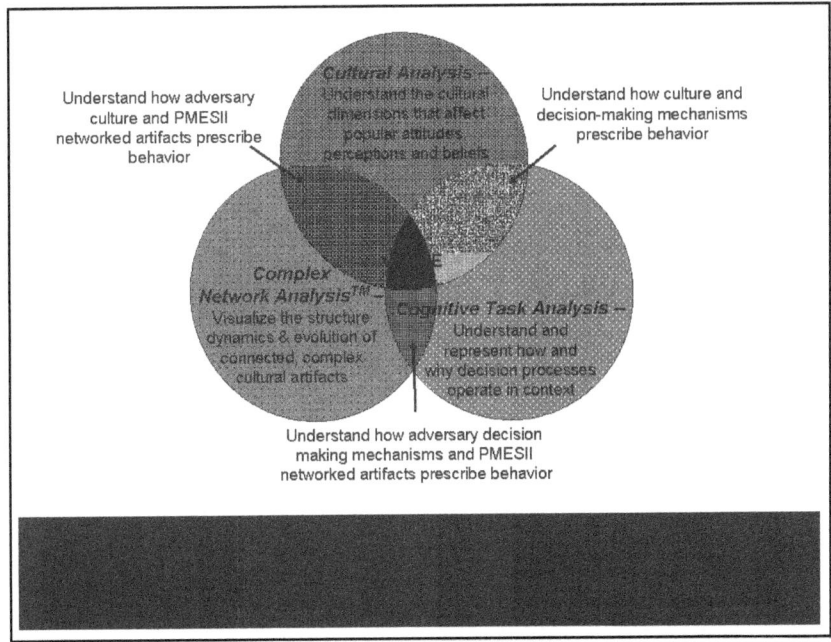

Understand how adversary culture and PMESII networked artifacts prescribe behavior

Cultural Analysis -- Understand the cultural dimensions that affect popular attitudes, perceptions, and beliefs

Understand how culture and decision-making mechanisms prescribe behavior

Complex Network Analysis[TM]—Visualize the structure, dynamics & evolution of connected, complex cultural artifacts

Cognitive Task Analysis -- Understand and represent how and why decision processes operate in context

Understand how adversary decision making mechanisms and PMESII networked artifacts prescribe behavior

FIGURE 3. CULTURAL-COGNITIVE SYSTEMS ANALYSIS (CCSA)[SM]

Additional Considerations (derived from Fisher's checklist).[61]

Fisher offers an approach for becoming more cross-culturally aware. While it was developed more for diplomats and foreign service officers than for members of the military, it provides a systematic way of diagnosing aspects of mind-sets in addition to our Cultural-Cognitive Dimensions. We, in the U.S., need to understand the particular contexts within which we and the adversary are operating. That means being able to answer such questions as:

- How do obvious differences in historical, geographical, or economic facts of life translate into special patterns of priorities and concerns? *For instance, clearly, at one time, Arabia was a center of learning and civilization. What must people feel to have lost that Golden Age, and to be perceived as inferior to the West? It must affect their attitudes and behavior towards us.*

- How does the context in which issues are presented affect the way they are perceived, or alter their dramatic or emotional impact? *Repetition, loud music, and somber voices heighten the drama of video footage. Routine military exercises may be perceived as saber rattling, when conducted near a country of interest.*

61. Fisher, *Mindsets*, 71-90.

- Do any unanticipated higher priority concerns or hidden agendas influence the perception of the issue? *These may range from misperceptions of motives to concern about the economic or "loss-of-face" costs of compliance with some agreement.*

We need to be aware of the knowledge and information base that we and our adversaries have.

- What knowledge or information base do people bring to the issue or event? *Has their education or their experience equipped them to understand the issues? Do they interpret the issues the same way?*

- What is the effect of new information such as that coming from the media or the internet? *Does it add to the knowledge base? Does it add to misperceptions? Is it propaganda? What impact is it likely to have on the public of our country and others?*

- What myths (including historical ones) are included in the information? *Much of the Middle East blames the West and/or Israel for its poverty and stagnation, for instance, and that myth permeates public perception of anything that we do.*

The Image Factor

- What images of the other side require consideration? *This is a difficult area, and we need to look carefully for the sources of the images. Some would say that we are so politically correct that we do not permit ourselves to recognize the violent aspects of Islam.*

- What images of us require consideration in the context of our goals? *The image of the Great Satan is still a pervasive one. And some people have not forgotten the Christian crusades. Handing out candy and toys to children in war zones projects a good image, but facing women and children while armed with rifles does not.*

- Which of our national self-images help explain reactions to issues or events? *We seem to see ourselves as the world's policeman—keeping order, maintaining the peace, being helpful. We are generous, and we find it difficult to imagine that people are willing to "bite the hand that feeds them."*

The final section of Fisher's checklist covers the cultural and social determinants of the dynamics between countries. He discusses mismatches in deep cultural beliefs, values, and assumptions. In fact, our Cultural-Cognitive Dimensions Analysis goes well beyond anything that Fisher discusses.

Conclusions

Future warfare is likely to include more action against terrorist groups and insurgents, more stability operations, and more operations in urban areas where our Armed Forces will have to deal with civilians as well as their targets. Even if we acquire a peer competitor, our knowledge of its culture is likely to be less comprehensive than we would like. This means that we need to think very carefully about the outcomes and results we want to achieve—from both traditional, kinetic warfare and from non-kinetic operations, including IO, PSYOPS, and SC—wherever in the world we may be operating. In addition, we need to ensure that the intelligence gathered by whatever "_INT" seems appropriate, is what we need. Without a thorough understanding of what we are up against in terms of cultures, mind-sets, perceptions, and truths, we will not be able to frame our real, desired outcomes or the strategies and tactics for achieving them.

Misperceptions or misattributions of motivations can produce unanticipated consequences of enormous magnitude.

I cannot emphasize enough the importance of this understanding of cultures. It is needed for our decision makers and planners; it is also needed by our soldiers, sailors, airmen, and Marines with their boots on the ground. In-depth education and training in cultures must become part of the curriculum for anyone going into combat overseas. This may require more time and effort than we have time for at present; it may even require different people from those we currently recruit and commission—but we need to start thinking along these lines.

APPENDIX

Applied Futures' / Cultural Dynamics' Social Models

Since the early '70's, Applied Futures and its sister company, Cultural Dynamics, have conducted studies of the future, including research into changing values, beliefs, and motivations. Our studies included a statistical model of the UK society in which Applied Futures' headquarters was located until 1993. During that same period, associates in other industrialized nations, including the United States, developed similar models.

The model we use is based on Abraham Maslow's theory of motivation,[62] and it has been validated by data from periodic surveys of random samples of the UK population conducted since 1973.[63] In the United States, similar surveys, conducted from 1968 to1986 showed very similar results.[64] Discussions with Professor Ronald Inglehart (University of Michigan) about two of his books provided further validation for this model.[65] Maslow's contention was that every individual has within his psychological framework a hierarchy of needs, as represented in figure A1. Shalom Schwartz has also produced a values-based model that has been tested in many different countries and which correlates well with our model.[66]

According to Maslow, the individual must satisfy, at least in part, the needs at one level of the hierarchy before he can be even conscious of those at the next level. Maslow saw this as a process of psychological development that takes place throughout life; so that by learning to satisfy the needs of progressive levels the individual can, in time, realize his full psychological potential. Maslow also made the distinction between "deficiency needs," which the individual perceives as some sort of lack, and "growth" needs, which are recognized as experiences needed for the realization of one's individual potential.[67] From our data and from discussions with colleagues around the

---

62. Maslow, *Motivation and Personality*.

63. Taylor Nelson, *Monitor Program Reports*, (Epsom, Surrey, 1973-1988); see also, MacNulty and Higgins, *Applied Futures Social Change Program Reports*.

64. SRI International, *VALS Program Reports*.

65. Inglehart, Culture Shift in Advanced Industrial Society, and Modernization and Postmodernization.

66. See in Shalom Schwartz, "Beyond individualism/collectivism" and in Shalom Schwartz, "Basic human values: an overview" (Jerusalem, Hebrew University 2005) < www.fmag.unict.it/Allegati/convegno%207-8-10-05/Schwartzpaper.pdf >.

67. Abraham Maslow "Deficiency motivation and growth motivation," *Nebraska Symposium on Motivation* (Nebraska, University of Nebraska, 1955).

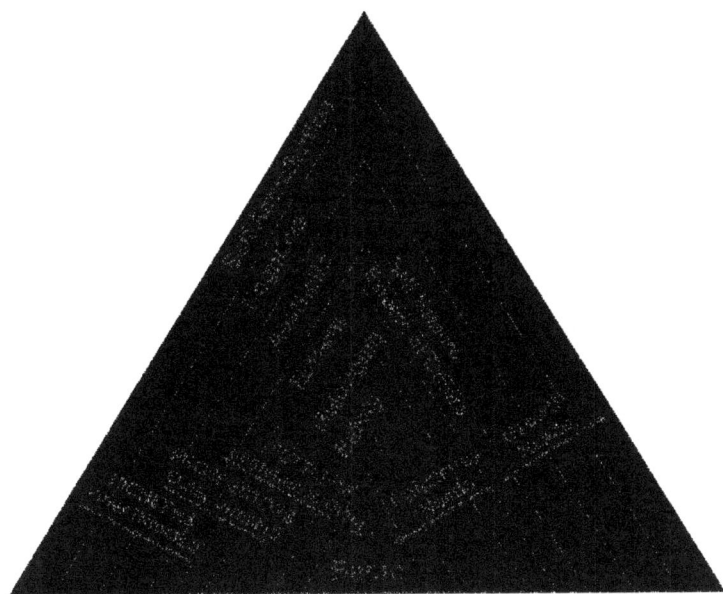

FIGURE A1. MASLOW'S HIERARCHY OF NEEDS

world, we believe that most of the industrialized countries are moving into the growth needs, while most Arab and Islamic countries are still in the deficiency needs. This is one of the major causes of Arab resentment.

The fundamental idea that underlies our social model is that every individual possesses a set of values and beliefs. These are based on the needs the individual perceives; and if they change at all, they change only slowly—as he works his way through the hierarchy of needs. These values and beliefs motivate almost everything the individual does. These long-term values (by long-term, we mean values that are held for a period of from five to twenty years) are manifested in the medium-term as attitudes and lifestyles, and in the short-term as behavior. By understanding their motivations, we can assess the likely attitudes and behavior of people to a much greater degree than we can by extrapolating behavioral data.

In addition to these social values being the most fundamental driving force for change, it is the very fact that they change slowly that makes them so important and valuable when taking a longer-term view; they are a relatively stable element in a world of "fast-changing" data. Our model proposes that by understanding the values of individuals in a society and the way those values are changing we can observe the developments taking place within that society, identify the way in which they are occurring, and (in a general way) assess the manner in which the society will behave.

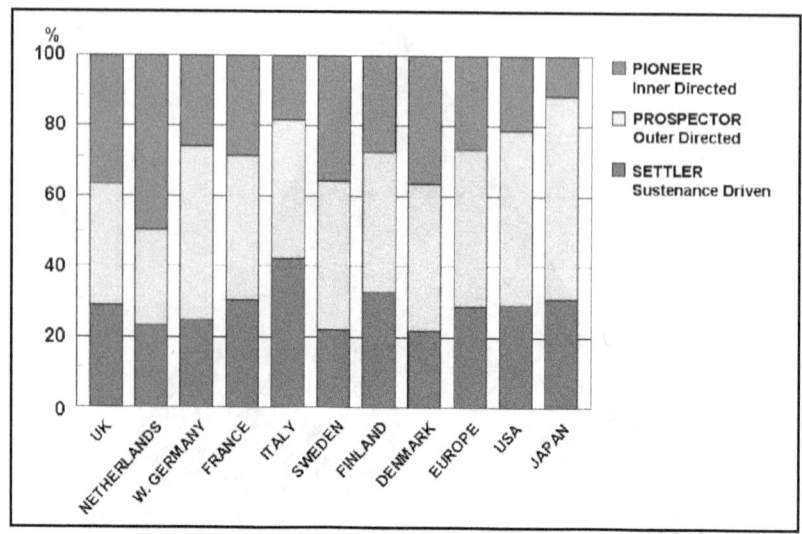

Figure A2. International Comparison of Values
(From MacNulty & Zetterberg for IBM, 1986)

We recognize three major groupings of social values within the United States, United Kingdom, other Western countries, and Japan. We refer to these groups as *Pioneers, Prospectors, and Settlers.* In each country we also have sub-groupings that give added refinement to the model. In the paragraphs below we describe briefly the principal identifying characteristics of each of these groups, set out what our understanding of their role in society is, and suggest how we expect them to influence the future. While Maslow saw that every individual has the potential for moving through the entire hierarchy of needs, he also recognized what he called "dominant motivation"—the tendency of an individual to have a "center of gravity" around which he operates. The three groups we have identified are named for the dominant motivation of the people who espouse them.

Examples of national groupings are given in figure A2, which was prepared for a project undertaken for IBM Europe.

Geert Hofstede, the Dutch anthropologist, has examined a number of countries in terms of a small number of dimensions.[68] We have plotted some of his work on to a map of our values, as shown in figure A3.

Pioneer Group
The Pioneer Group is so named because its members derive their sense of personal direction, their personal rewards, and their criteria for success

---

68. Hofstede, *Cultures and Organizations.*

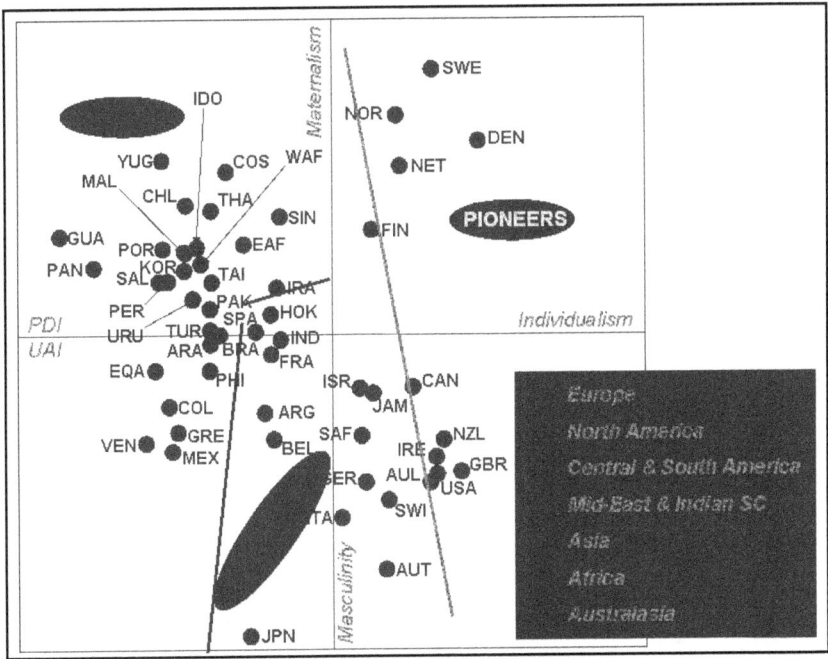

FIGURE A3. GEERT HOFSTEDE'S CULTURAL MODEL
(WITH FACTOR ANALYSIS BY CULTURAL DYNAMICS)

from exploring the bounds—physical, mental, and emotional. The standards by which Pioneers measure themselves, and the world, tend not to be the materialistic standards of wealth, social class, income, status, or possession; but rather they are standards involving such things as integrity, honesty, quality, and appropriateness to the situation. Pioneers are the most psychologically mature of the groups, yet they are still seeking greater maturity. Although most are not anti-materialistic, they consider people rather than things to be of paramount importance; therefore, they tend to see people in ways which have far greater human significance than the social role of membership in a class or the economic role of producer/consumer.

Their introspection is a characteristic that makes these people very easy to misunderstand; and it is important not to confuse Pioneer introspection with introversion or self-centeredness. While they do try to maximize their own, individual, potential, they generally seek to do that in a way that is not exploitive of others.

Pioneers are difficult to observe because the thing that distinguishes them from the rest of the population is their *motivation* rather than their *behavior*; and as a result of this, the media has real difficulty in presenting Pioneers.

Pioneers tend to be self confident; and although they are by no means anti-social, they do not feel obliged to conform to stereotyped social "norms."

It seems likely that it is the combination of their self confidence and their inner sense of what is important in their lives that gives the Pioneer group its significant role as "trendsetters" in society. During the past thirty years, almost every major trend in Western societies has been started by this group, although the trends have then been picked up and driven, as fashion, by Prospectors. In almost every area, the Pioneer group has influence out of all proportion to its size. Indeed, the *1998 Yankelovich Monitor,* in describing "America's New Agenda" suggested that Generation X had adopted many earlier Pioneer approaches to life. The Pioneer group has been growing slowly for the past thirty years; and although the growth rate has slowed recently, we expect it to continue to grow slowly in most industrialized countries. Indeed, in the United States it has grown from 22% twenty years ago to 37% of the population today.[69] However, it is possible that, if industrialized countries experience significant immigration, the proportion of Pioneers may decline. In any case, we expect that they will maintain significant influence.

With respect to the "bad" side of Pioneers, we should note that they are, at heart, idealists. From what we have seen through the media of Osama bin Laden, he is probably a Pioneer whose idealism was thwarted, and hence became an ideology

Prospector Group

In contrast to inner-directed people, Prospectors rely heavily on external indicators of their own self worth. To put it another way, a Prospector's concept of himself depends upon his being able to compare himself with others; his self esteem depends upon finding himself to be "better off"—usually in some materialistic way. Since they are, especially in the West, characterized by the idea that "you are what you consume, they form an easily visible group."[70] Display, and particularly the display of possessions and "badges," is a necessary element in establishing their place in society. This shows clearly in their homes, which tend to be neat, tidy, and well organized, with their most prestigious possessions, particularly consumer goods, openly exhibited. Display can also be seen in evidence of recent exotic vacations or activities, or in the use of high quality fitness and sporting gear. In Latin

---

69. In 1986, Hans Zetterberg from SIFO in Sweden conducted research for IBM that included surveys of the populations of a number of countries, including the United States. The author contributed to that project. In 2006 the Canadian Social Research company, Environics, used those questions on a survey they did in the United States. These percentages are based on that survey.

70. Faith Popcorn, *The Popcorn Report,* (New York, Doubleday, 1991)

America and some other countries, where success comes from social status rather than material things, Prospectors will seek social success and the power that accompanies it.

As the notation on Maslow's diagram indicates, Prospector needs are centered on esteem. Therefore, at work the Prospector person is conscious of, and seeks actively to acquire, status and the symbols related to it. Such people are very much at home in structured, hierarchical organizations in which they can establish their position clearly and then display their position and measure their progress relative to others. In identifying themselves with a peer group in this way, Prospectors automatically judge themselves to be up to the group's level, and they generally use the group as the source of the standards by which they judge their world. The people in this group are of vital social importance; they are the dynamo, the energy source, in our society. They are the ones who feel the need to compete, who need to prove themselves against the opposition, who have the drive to win at virtually any cost. This Prospector energy is essential to business as it operates today and to society in general if it is not to stagnate. This group grew very quickly during the '80's, declined a little during the early '90s, and now appears to be bouncing back. In the United States, this group is 50% of the population. For the next decade, at least, we expect it to continue to grow in most of the industrialized and in the rapidly developing countries as long as their economies are viable.

Settler Group

Everywhere we look in Western industrial society, the two groups that we have just considered are growing at the expense of a third group, which we call the Settlers. This pattern has been a consistent trend for some time. Because of its declining size, we consider the direct impact of the Settler group on the long-term future of Western countries to be relatively small. However, they are influential at the moment, and to neglect them would be to miss the essential role they will play in influencing the future. Indeed, if the industrialized countries experience significant immigration from the developing world, this group may increase in size, although that has not been the case in the United States, where it has declined to 13% of the population.

Settler needs are deficiency needs, and the distinguishing characteristic of the Settler is a desire to "hold what you've got." This orientation tends to make them form homogeneous groups with well-defined characteristics and relatively impermeable boundaries. The typical picture that this idea brings to mind is the tightly knit, clannish, working-class community. A little reflection will indicate that these characteristics also describe a good many company directors of the "old school," a lot of the traditional "professions" including a university faculty, not to mention a good part of the hereditary peerage in

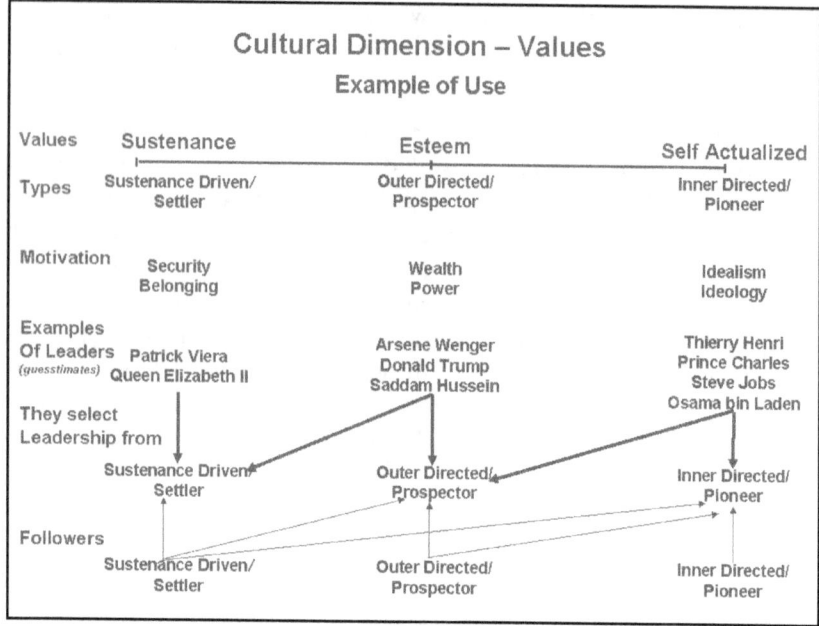

FIGURE A4. CULTURAL DIMENSION – VALUES – EXAMPLE OF USE

the UK. In fact, we find that the Settler groups include a substantial number of people from all these conventional classifications, and the thing that they have in common is that they resist change. Not only do they hold on to their possessions, but to their institutions as well.

The majority of the populations of developing countries are likely to be Settler, not just because some of them are poor, but because of close-knit tribal values and a political power structure that works to keep them dependent. With a few exceptions, the leaders of countries in the developing world seem to be Prospectors. They appear to be motivated by the need to acquire and maintain power—and the wealth that goes with it—for themselves and often for their families. They do not seem interested in wealth for everyone.

Comment

It is important to understand that these statements of function are not value judgments. Nor is it sufficient to think of these groups as being in competition one with another, although the Settlers, particularly, might be prone to that view. Rather it is necessary to take a system's view of society and understand that each of these functions is both essential and beneficial to the overall well-being of society. We are looking at a symbiosis in which Settler inertia keeps Prospector enthusiasm in check, while the Prospectors, in turn, provide a focus that prevents a Pioneer evaporation into personal space. Conversely, the Pioneers constantly provide the Prospectors with new ideas and opportunities, while the Prospectors provide the energy to drive businesses that service their material needs and those of the Settlers. What we have here, in fact, is a psychological view of the great dynamic of society; it is the interplay of these three forces that really keeps the wheels moving throughout the industrialized world and which, in our view, provides the major driving force for change.

To illustrate very simply how this model can be used, let us look at some of the characteristics and some personalities as shown in figure A4. With the exception of the Arsenal Soccer team members (Veira, Wenger, and Henri), on whom we have data, the rest are guesstimates based on what we know about them from the media. However, such guesstimates can often be correct.

However, the way in which they select their leaders and followers is generally correct. Since we know their values, we can understand how to motivate the different groups to action, as we have done for commercial organizations, their employees, and their customers.

SELECTED REFERENCES

Ajami, Fouad. *The Dream Palace of the Arabs.* New York: Pantheon Books, 1998.

Aspin, Les. "Misreading Intelligence," *Foreign Policy* 43 (Summer 1981) 166-72.

Cook, David. "The recovery of radical Islam in the wake of the defeat of the Taliban," in *Terrorism and Political Violence,* 15, 1, Spring 2003.

de Rivera, Joseph H. *The Psychological Dimension of Foreign Policy.* Columbus, OH: Charles E. Merrill Publishing Company, 1968.

Fisher, Glen. *Mindsets.* Yarmouth ME: Intercultural Press, Inc, 1988.

Garcia, Jorge J. E. *Old Wine in New Skins.* Milwaukee: Marquette University Press, 2003.

Goldberg, Bernard. *Bias: a CBS Insider Exposes how the Media Distort the News.* Washington DC, Regnery Publishing, Inc., 2002.

Hofstede, Geert. *Cultures and Organizations.* London: McGraw Hill, 1991.

Hoover, Kenneth. *The Power of Identity: Politics in a New Key.* Chatham, NJ: Chatham House Publishers, 1997.

Inglehart, Ronald. *Culture Shift in Advanced Industrial Society.* Princeton, NJ: Princeton University Press, 1990.

———. *Modernization and Postmodernization.* Princeton, NJ: Princeton, University Press, 1997.

Kimmage, Daniel, and Kathleen Ridolfo. *Iraqi Insurgent Media: the War of Images and Ideas.* Available on the Radio Free Europe/Radio Liberty website at < realaudio.rferl.org/online/OLPDFfiles/insurgent.pdf >.

Klein, Helen A. "Cognition in natural settings: the cultural lens model" in Kaplan (Ed). *Cultural Ergonomics, Advances in Human Performance and Cognitive Engineering.* Oxford: Elsevier Press, 2004.

Lakoff, George, and Mark Johnson. *Metaphors We Live by.* Chicago: University of Chicago, 1980.

Lewis, Bernard. *Islam and the West.* Oxford: Oxford University Press, 1993.

MacNulty, Christine, and Leslie Higgins. *Applied Futures Social Change Program Reports*. London: 1988-1993.

MacNulty, Christine, William Ross, and Jeffrey Cares. "Cultural-Cognitive Systems Analysis (CCSA)^{SM}.

MacNulty, W. Kirk. "The paradigm perspective," *Futures Research Quarterly*, 5, 3, 1989 35-54.

Maslow, Abraham. "Deficiency motivation and growth motivation" in *Nebraska Symposium on Motivation*. Nebraska: University of Nebraska, 1955.

———. *Motivation and Personality*. New York: Harper Row, 1954.

Maznevski, Martha, et al. "Cultural dimensions at the individual level of analysis" *International Journal of Cross-Cultural Management*, 2002, 2.

McConnell, Michael. *Stepping Over: Personal Encounters with Young Extremists*. New York: Reader's Digest Press, 1983.

Nelson, Taylor. *Monitor Program Reports*. Epsom, Surrey: 1973-1988.

Nisbett, Richard. *The Geography of Thought: How Asians and Westerners Think Differently and Why*. New York: The Free Press, 2003.

Patai, Raphael. *The Arab mind*. New York: Hatherleigh Press, Revised edition, 2002.

Peng, K. and R. Nisbett. "Culture, dialectics and reasoning about contradiction," *American Psychologist*, 54, 741-754.

Popcorn, Faith. *The Popcorn Report*. New York: Doubleday, 1991.

Pryce-Jones, David. *The Closed Circle: an Interpretation of the Arabs*. Chicago: Ivan R. Dee, new edition, 2002.

Rosenthal, Robert. Biography. 2007, University of California, Riverside, June 6, 2007 < www.facultydirectory.ucr >.

Said, Edward W. *Orientalism*. New York: Random House, 1978.

Schwartz, Shalom. "Basic human values: an overview." Jerusalem: Hebrew University 2005) < www.fmag.unict.it/Allegati/convegno%207-8-10-05/Schwartzpaper.pdf >.

———. "Beyond individualism/collectivism: new cultural dimensions of values" in Kim et al (Eds) *Individualism and Collectivism: Theory, Methods and Applications*, (Newbury Park, CA, Sage, 1994.

Segall, Marshall H. et al. *The Influence of Culture on Visual Perception.* New York: Bobbs-Merrill, 1966.

Stossel, John. *Myths, Lies, and Downright Stupidity.* New York, Hyperion, 2006.

Tart, Charles. *Waking up: Overcoming the Obstacles to Human Potential.* Boston, MA: Shambala Publications, 1986

SRI International. *VALS Program Reports.* Menlo Park, CA: 1978-1981.